YOUR KNOWLEDGE HAS VALUE

- We will publish your bachelor's and master's thesis, essays and papers

- Your own eBook and book - sold worldwide in all relevant shops

- Earn money with each sale

Upload your text at www.GRIN.com
and publish for free

Bibliographic information published by the German National Library:

The German National Library lists this publication in the National Bibliography; detailed bibliographic data are available on the Internet at http://dnb.dnb.de .

This book is copyright material and must not be copied, reproduced, transferred, distributed, leased, licensed or publicly performed or used in any way except as specifically permitted in writing by the publishers, as allowed under the terms and conditions under which it was purchased or as strictly permitted by applicable copyright law. Any unauthorized distribution or use of this text may be a direct infringement of the author s and publisher s rights and those responsible may be liable in law accordingly.

Imprint:

Copyright © 2017 GRIN Verlag, Open Publishing GmbH
Print and binding: Books on Demand GmbH, Norderstedt Germany
ISBN: 9783668521643

This book at GRIN:

http://www.grin.com/en/e-book/373965/mechanical-metallurgical-and-fatigue-properties-of-friction-stir-welded

Akshansh Mishra, A Razal Rose, Saravanan .M, Anand Singh

Mechanical, Metallurgical and Fatigue Properties of Friction Stir Welded and Tungsten Inert Gas Welded AA6061-T6 Aluminium Alloys

A Comparative Study

GRIN Publishing

GRIN - Your knowledge has value

Since its foundation in 1998, GRIN has specialized in publishing academic texts by students, college teachers and other academics as e-book and printed book. The website www.grin.com is an ideal platform for presenting term papers, final papers, scientific essays, dissertations and specialist books.

Visit us on the internet:

http://www.grin.com/

http://www.facebook.com/grincom

http://www.twitter.com/grin_com

Comparative study of mechanical, metallurgical and fatigue properties of friction stir welded and TIG welded AA6061-T6 aluminium alloys

Akshansh Mishra, Dr. A Razal Rose, Saravanan .M, Anand Singh

Department of Mechanical Engineering, SRM University, Kattangulathur, Chennai

Abstract: In this study, 6061-T6 aluminium alloy plates in 4mm thickness that are particularly used for aerospace and automotive industries were welded using Tungsten Inert Gas(TIG) welding and Friction Stir Welding(FSW) methods as similar joints with one side pass with parameters of varying tool rotation, weld speed and 2.3 degree tool tilt angle. Tensile tests results showed high yield stress values for FSW joints. The weld zones cross sections were analysed with light optical microscopy (LOM). Micro Vickers hardness test gave the required result for FSWed and TIG welded AA6061-T6 aluminium alloy plates. While fatigue test results showed all similar welded joints have fatigue strength close to each other.

Keywords: Weld Zones, Friction stir welding, Metallurgical Test, Mechanical Test

1. Introduction

This paper deals with the mechanical, microstructural and fatigue analysis of welded joints made by the Friction Stir welding process and Tungsten Inert Gas (TIG) Welding process. We have general welded similar plates of dimensions 150mm×100mm×4mm of AA6061-T6 aluminium alloy. Low weight, good weldability and good mechanical properties make aluminium alloys commonly used engineering materials. However, it was found that aluminium was difficult to weld with conventional fusion welding process due to formation of intermetallic compounds between the joints which results in poor strength. Friction Stir Welding is a solid state joining process developed in 1990 and is nowadays frequently used to join aluminium alloys. The FSW process is fast and it can be automated easily. This results decrease in manufacturing cost and production time. While Tungsten Inert Gas welding which is also known as Gas Tungsten Arc Welding is one of the popular welding process usually employed in the fabrication of thin structures. Previous study indicate that the fatigue properties of TIG welds differs from those of conventionally welded. Due to this level of fatigue strength and also the slope of S-N curves for FSWed is different compared to existing fatigue design standards for arc welding. In transportation and under varying load conditions fatigue failure is an important issue. There are many factors that make the weld critical under fatigue loading conditions. For instance, stress concentrations such as weld toe and weld root, residual stresses, unfavourable mechanical properties of the weld nugget and potential defects in the weld are major causes of weld failure in service. Weld failure leads to loss of lives and substantial costs each year all over the world. Fatigue analysis helps to identify how repetitive load cycles cause structural failures. It helps us to identify failures in components subjected to stress less than yield and do not experience plastic deformation and have relatively long lives. M. Czechowski - *Low-cycle fatigue of friction stir welded Al–Mg alloys*, the author in this paper [1] proposes the following alloys EN-AW 5058 H321 and

EN-AW 5059 H321 (Alustar) were welded by FSW (friction stir welding) method. The FSW welds showed better properties in comparison to the joints welded by the MIG method. The test of microstructure proved the proper structure of the weld which consisted of following: welded nugget, thermo-mechanically affected zone (TMAZ), heat affected zone (HAZ) and unaffected material. D.R. Ni, D.L. Chen, J. Yang, Z.Y. Ma - *Low cycle fatigue properties of friction stir welded joints of a semi-solid processed AZ91D magnesium alloy*, the authors in this paper [2] proposes a semi-solid processed (thixomolded) Mg–9Al–1Zr magnesium alloy (AZ91D) was subjected to friction stir welding (FSW), aiming at evaluating the weldability and fatigue property of the FSW joint. Microstructure analysis showed that a recystallized fine-grained microstructure was generated in the nugget zone (NZ) after FSW. The yield strength, ultimate tensile strength, and elongation of the FSW joint were obtained to be 192 MPa, 245 MPa, and 7.6%, respectively. Low-cycle fatigue tests showed that the FSW joint had a fatigue life fairly close to that of the BM, which could be well described by the Basquin and Coffin-Manson equations. G. Padmanaban, V Balasubramanian, G. Madhusudhan Reddy - *Fatigue crack growth behaviour of pulsed current gas tungsten arc, friction stir and laser beam welded AZ31B magnesium alloy joints*. The author's in this paper [3] proposes the laser beam welded joints offered better resistance against the growing fatigue cracks compared to friction stir welded and pulsed current gas tungsten arc welded AZ31B magnesium alloy joints. The formation of very fine grains in weld region, higher fusion zone hardness, uniformly distributed fine precipitates and favourable residual stress field of the weld region are the main reasons for superior fatigue performance of laser beam welded joints of AZ31B magnesium alloy. Michael Besel, Yasuko Besel Ulises Alfaro Mercado, Toshifumi Kakiuchi, Yoshihiko Uematsu - *Fatigue behaviour of friction stir welded Al–Mg–Sc alloy*, the authors in this paper [4] proposes the effect of Friction Stir Welding on the fatigue behaviour of Al–Mg–Sc alloy has been studied. To reveal the influence of the welding parameters, different travel speeds of the welding tool have been used to provide weld seams with varying microstructural features. Crack initiation as well as crack propagation behaviour under fatigue loading has been investigated with respect to the local microstructure at the crack initiation sites and along the crack path. L. Boni, A. Lanciotti, C. Polese - *"Size effect" in the fatigue behaviour of Friction Stir Welded plates*, the authors in this paper [5] proposes a Comparative fatigue tests were carried out on Friction Stir Welded specimens of a 2195-T8 aluminium– lithium alloy that differed significantly in width. The width of the larger specimens was over thirteen times greater than that of the small specimens. Fatigue results showed a clear "size effect", i.e. fatigue life of large specimens was about 40% of the corresponding value of small specimens. The Equivalent Initial Flaw Size methodology was adopted to correlate the two sets of results. Fatigue crack initiation life was disregarded with respect to crack propagation life, and fatigue life was evaluated only as propagation of a small pre-existing defect. Athanasios Toumpis, Alexander Galloway, Lars Molter, Helena Polezhayeva ,*Systematic investigation of the fatigue performance of a friction stir welded low alloy steel*, the authors in this paper [13] proposes A comprehensive fatigue performance assessment of friction stir welded DH36 steel has been undertaken to address the relevant knowledge gap for this process on low alloy steel. A detailed set of experimental procedures specific to friction stir welding has been put forward, and the consequent study extensively examined the weld microstructure and hardness in support of the tensile and fatigue testing. The effect of varying welding parameters was also investigated. Micro structural observations have been correlated to the weldments fatigue behaviour.

During recent years several investigations have been made of fatigue properties of friction stir welded joints. The great majority of available data from the fatigue analysis of friction stir welded joints are concerned with uniaxial loading conditions for a simple geometry. In uniaxial loading nominal stress is normally used as reference stress and it is easy to determine. However, fatigue failure is a highly localized phenomenon in engineering components and determining the nominal stress is not always possible due to the complexity of structures and presence of stress concentrators such as notches and cracks in which many approaches based on local parameters.

2. Materials and experimental procedure

In this study, 6061-T6 aluminium alloys were used as base metals. Aluminium alloy plates were machined to required dimensions for butt welding. For Friction Stir Welding H13 tool steel with chemical composition 0.406% C, 1.096% Si, 0.443% Mn, 4.952% Cr, 1.251% Mo, 0.183% V with given dimensions was used as weld tool as shown in Fig.1. Varying welding parameters like tool rotation speed, welding speed and feed depth were used as shown in Table 1. Similar joints of 6061-6061 aluminium alloys were fabricated using these parameters. While during TIG welding, the parameters used for welding similar joints 6061-6061 aluminium alloys, power source was AC at a frequency of 50Hz with current and voltage value of 1600A and 40V. The filler material used in TIG welding was ER-4043. This filler material is Silica rich compound which inhibits the formation of intermetallic compounds which results in poor weld quality. In order to determine microstructure properties of these joints, the specimens were cross- sectioned perpendicular to the weld interface using of low speed band saw. Using emery papers of grade 600, 1200, 1500, 2000 the cross section of these joints were metallographically polished and Keller etchant was used for LOM examination. These specimens were used for microhardness assessment under 100g load after microstructure test. For fatigue test, standard specimen was prepared as shown in Fig. 2. The specimen was subjected to low cycle fatigue test.

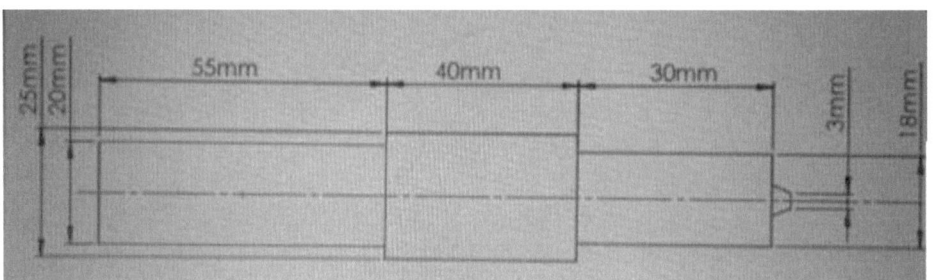

Fig.1 FSW Tool geometry (without threaded)

Table 1 FSW parameters for joining AA6061-T6 similar plates

Welding Run no	Rotational Speed(rpm)	Welding Speed(mm/m)	Tool Diameter(mm)	Depth of Plunging(mm)
1	1600	40	6	3.75
2	1600	40	6	3.60
3	1600	35	6	3.72
4	1750	45	6	3.65
5	1800	50	6	3.70

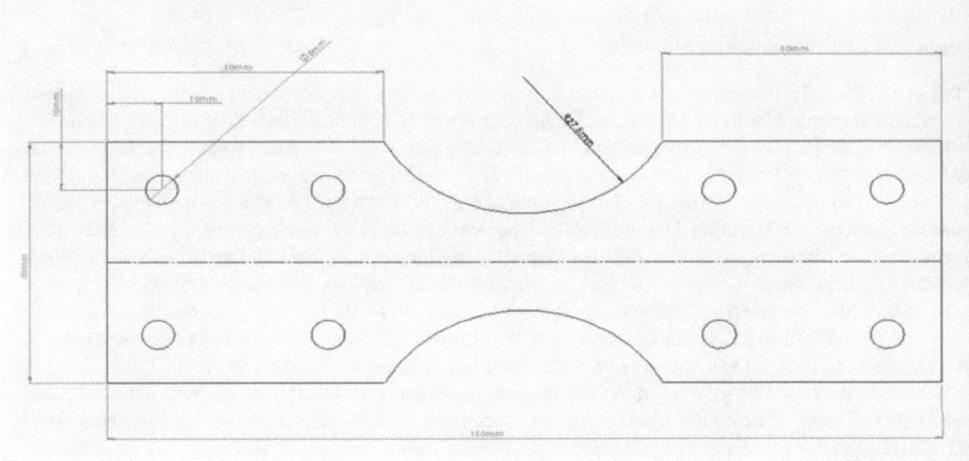

Fig 2. Fatigue test specimen

3. Results and discussions

3.1 *Microstructural behaviour*

By using LOM, the microstructural behaviour of FSW and TIG welded joints of aluminium alloy were studied. Images of different types of Friction Stir weld zones evaluated are shown in Fig. 3.

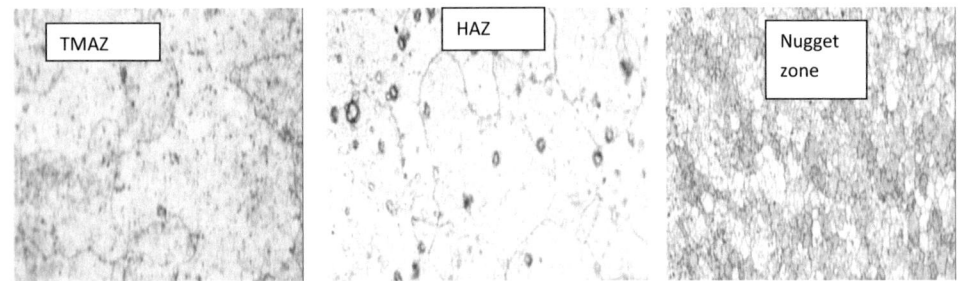

Fig 3 Different weld zones in FSWed joints

Basically, there are four types of zones observed in FSW joints: Thermo-mechanical Heat Affected Zone(TMAZ), Heat Affected Zone(HAZ), Nugget Zone(NZ), Base Metal Zone. From the obtained microstructure it is obseved that NZ has equi-axed grains. This zone is also called Stir Zone. A unique feature of the stir zone is the common occurrence of several concentric rings which has been referred to as an "onion-ring" structure. It is obseved that the size of grains in NZ is smaller than in the other zones. While in TMAZ, unlike the stir zone the microstructure is recognizably that of the parent material. Also the strain and temperature are lower and the effect of welding on the microstructure is correspondingly smaller. In HAZ, the temperatures are lower than those in the TMAZ but may still have a significant effect if the microstructure is thermally unstable. In fact, in age-hardened aluminium alloys this region commonly exhibits the poorest mechanical properties. Now coming to the obsevred microstructure of TIG welded joints, its shown in the Fig. 4.

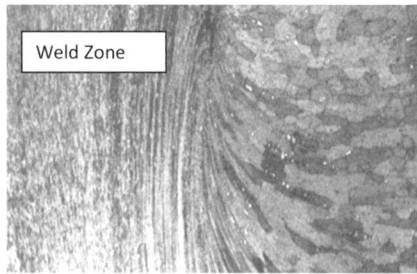

Fig.4 Weld Zones in TIG welded similar AA6061-T6 alloys

From the microstructure it is observed that in the weld zone there is formation of intermetallic dendtritic structure while in the HAZ there is formation of intermetalic granular structures.

3.2 *Micro hardness Test*

The Vickers method is based on an optical measurement system. The Micro hardness test procedure, ASTM E-384, specifies a range of light loads using a diamond indenter to make an indentation which is measured and converted to a hardness value. The datas for microhardness were obtained with 100gm load. A square base pyramid shaped diamond is used for testing in the Vickers scale. The hardness value for FSW is given in the form of graph as shown in Fig.5. The hardness value was taken from distance of every 2mm from weld joint.

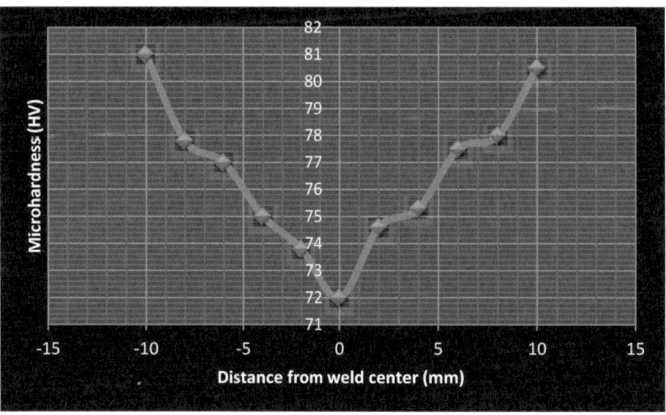

Fig. 5 Microhardness for FSW AA6061-T6 Joints

It is obseved that Base metal zone has more hardness value while Nugget zone has less hardness value compared to TMAZ and HAZ. According to Hall- Petch equation the hardness in Nugget zone should be higher but the main resaon for its less hardness is melting of precipates due to supply of heat. It is also observed that the graph is not symmetrical. The main reason is improper flow of material during joining process. The hardness value of TIG welded similar AA6061-T6 is tabulated in Table 2.

Table 2. Hardness value of TIG welded similar joints in HV

Current (A)	Weld Zone	Fusion Zone	Heat Affected Zone
160	71.06	76.06	80.23

3.3 *Tensile Test*

Tensile properties of both FSWed and TIG welded joints are given in Table 3 and Table 4. Tensile test results for FSW joints were above values results has been obtained from previous studies. It was observed that the failure occurs at the zone which has less hardness i.e in the weld zone. This was due to majority strength loss in weldments in strain-hardened due to the recrystallization process.

Table 3. Tensile test for FSWed joints

FSW joint	Area mm^2	Ultimate Load (KN)	Gauge Length (mm)	Ultimate Tensile Strength (MPa)	0.2% yield Strength (MPa)	% Elongation
Specimen 1	58.18	16.98	50	293	168	8.8
Specimen 2	57.87	16.88	50	290.65	164	8.3

Table 4. Tensile test for TIG welded joints

TIG joint	Area mm^2	Ultimate Load (KN)	Gauge Length (mm)	Ultimate Tensile Strength (MPa)	0.2% yield Strength (MPa)	% Elongation
Specimen 1	49.6	12.7	50	260	157	5.08
Specimen 2	51.1	12.5	50	250	160	5.12

The FSWed joints has more yield stress compared to TIG welded joints. And also FSWed joints has more elongation than TIG welded joints.

3.4 *Fatigue behavior*

For carrying out fatigue test, mount the test specimen given. Paste the strain sensor properly over the test specimen. Plug in the power cord to a single phase mains supply. Switch on the power to the equipment. The observations made during fatigue test are shown in Table 5 and Table 6.

Table 5. Fatigue test for FSWed joints

Frequency (HZ)	Motor Speed (rpm)	Movement (mm)	No. of cycle	Stress (MPa)
12.49	749	0.215	2609	136
12.87	773	0.217	3479	132
13.23	798	0.217	4063	127
13.81	829	0.223	4751	123
14.01	851	0.265	5628	119

Table 6. Fatigue test for TIG welded joints

Frequency (HZ)	Motor Speed (rpm)	Movement (mm)	No. of cycle	Stress (MPa)
10.75	645	0.425	3364	125
11.11	666	0.427	3915	122
12.56	719	0.432	4582	116
13.72	785	0.444	5145	113
15.34	920	0.478	5936	109

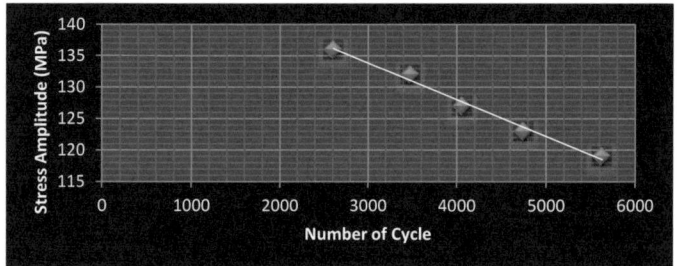

Fig. 6 SN curve for FSWed joints

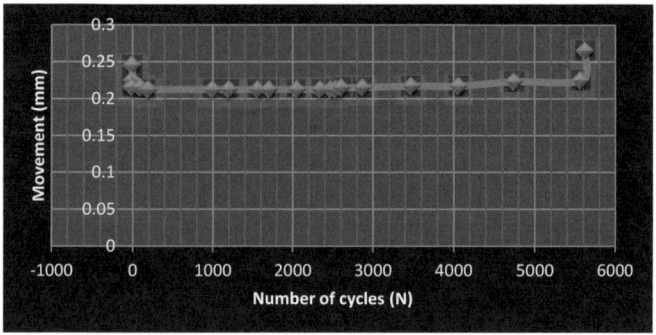

Fig. 7 Movement vs No. of cycles of FSWed joints

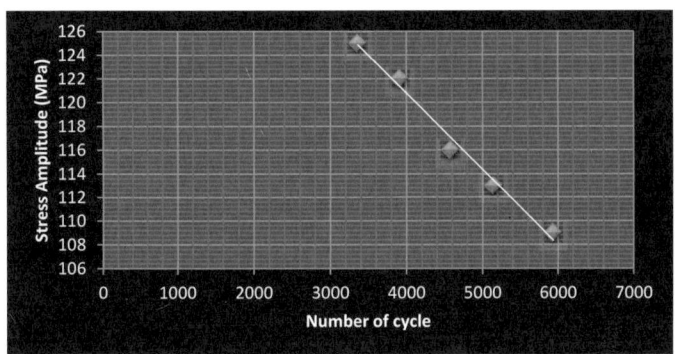

Fig.8 SN Curve for TIG welded joints

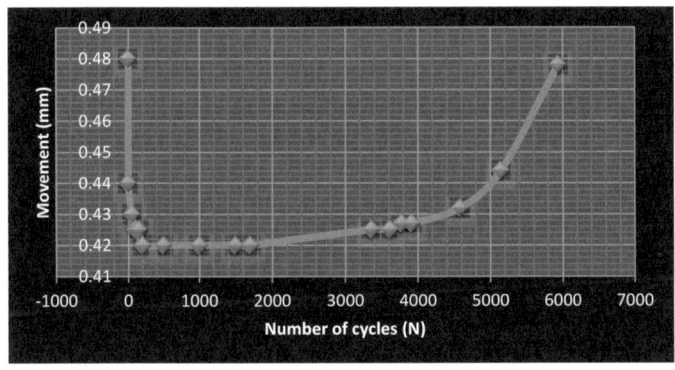

Fig. 8 Movement vs No. of cycles for TIG Welded joints

4. Conclusions

In this study, 4mm thick AA6061-T6 aluminium alloys that used widely in aerospace and automobiles industries were welded successfully using friction stir welding and TIG welding for similar alloy pair. The following conclusions were drawn:

- The formation of fine equiaxed grains and uniformly distributed very fine strengthening precipitates in the weld region is the reason for superior tensile property of FSW joints compared to TIG joints.
- Tensile test shows that FSW joints have higher strength and higher ductility compared to TIG joints.
- Heat Affected Zone is narrower than TIG welding process. FSW requires less pre operation than TIG welding process.
- From Industrial perspectives, FSW is very competitive because it saves energy due to less heat input.
- FSW prevents joints from fusion related defects.
- FSW has better strength than TIG welding process.
- Aluminium AA6061-T6 alloy welded by FSW method shows greater fatigue life in comparison to same alloy welded by TIG welding method.
- Fatigue failure was generally occurred in the zone among flow arm, weld nugget and Thermo Mechanically Affected Zone (TMAZ).
- Fatigue limits of all joint types were close to each other.

References

1. M. Czechowski (2005) *'Low-cycle fatigue of friction stir welded Al–Mg alloys'* Journalof Materials Processing Technology, Vol.164-165, pp.1001-1006.

2. D.R. Ni, D.L. Chen, J. Yang, Z.Y. Ma (2014) *'Low cycle fatigue properties of friction stir welded joints of a semi-solid processed AZ91D magnesium alloy'* Journal of Materials and Design, Vol.56, pp.1-8.

3. G. Padmanaban, V. Balasubramanian, G. Madhusudhan Reddy(2011) *'Fatigue crack growth behaviour of pulsed current gas tungsten arc, friction stir and laser beam welded AZ31B magnesium alloy joints'* Journal of Materials Processing Technology, Vol.211, pp.1224-1233

4. Michael Besel, Yasuko Besel, Ulises Alfaro Mercado, Toshifumi Kakiuchi, Yoshihiko Uematsu (2015) *'Fatigue behavior of friction stir welded Al–Mg–Sc alloy'* International Journal of Fatigue, Vol.77, pp.1-11

5. L. Boni, A. Lanciotti, C. Polese (2015)*'Size effect in the fatigue behaviour of Friction Stir Welded plates'* International Journal of Fatigue, Vol.80, pp.238-245

6. V.X. Tran, J. Pan, T. Pan (2010) *'Fatigue behaviour of spot friction welds in lap-shear and cross-tension specimens of dissimilar aluminium sheets'* International Journal of Fatigue, Vol.32, pp.1022-1041

7. D.R. Ni, D.L. Chen, B.L. Xiao, D. Wang, Z.Y. Ma (2013) *'Residual stresses and high cycle fatigue properties of friction stir welded SiCp/AA2009 composites'* International Journal of Fatigue, Vol.55, pp.64-73

8. H.M. Rao, J.B. Jordon, B. Ghaffari, X. Su, A.K. Khosrovaneh, M.E. Barkey, W. Yuan, M. Guo (2016) *'Fatigue and fracture of friction stir linear welded dissimilar aluminum-to-magnesium alloys'* International Journal of Fatigue,Vol.82, pp.737-747

9. Z. Barsoum, M. Khurshid, I. Barsoum (2012) *'Fatigue strength evaluation of friction stir welded aluminium joints using the nominal and notch stress concepts'* Materials and Design, Vol.41, pp.231-238

10. Soran Hassanifard, Masoud Mohammadpour, Hossein Ahmadi Rashid (2014) *'A novel method for improving fatigue life of friction stir spot welded joints' using localized plasticity'* Materials and Design, Vol.53, pp.962-971

11. Beytullah Gungor, Erdinc Kaluc, Emel Taban, Aydin Sik (2014) *'Mechanical, fatigue and microstructural properties of friction stir welded 5083-H111 and 6082-T651 aluminum alloys'* Materials and Design, Vol.56, pp.84-90

12. Yong Zhao, Zhengping Lu, Keng Yan, Linzhao Huang (2015) *'Microstructural characterizations and mechanical properties in underwater friction stir welding of aluminum and magnesium dissimilar alloys'* Materials and Design,Vol.65, pp.675-681
13. Athanasios Toumpis, Alexander Galloway, Lars Molter, Helena Polezhayeva (2015) *'Systematic investigation of the fatigue performance of a friction stir welded low alloy steel'* Materials and Design, Vol.80, pp.16-128
14. Banglong Fu, Guoliang Qin, Fei Li, Xiangmeng Meng, Jianzhong Zhang, Chuansong Wu (2015) *'Friction stir welding process of dissimilar metals of 6061-T6aluminum alloy to AZ31B magnesium alloy'* Journal of Material Processing Technology, Vol.218, pp.38-47
15. H. Wohlfahrt, Th. Nitschke-Pagel, W. Zinn (2004) *'Optimization of the fatigue behaviour of welded joints by means of shot peening - a comparison of results on steel and aluminium joints'* Fatigue Fracture Engineering Materials Structure, Vol.27, pp.785-795
16. G. M. Xie, Z. Y. Ma2 and L. Geng1 (2015) *'Effects of Friction Stir Welding Parameters on Microstructures and Mechanical Properties of Brass Joints'* Materials Transactions, Vol.49, No.7, pp.1698-1701

YOUR KNOWLEDGE HAS VALUE

- We will publish your bachelor's and
  master's thesis, essays and papers

- Your own eBook and book -
  sold worldwide in all relevant shops

- Earn money with each sale

Upload your text at www.GRIN.com
and publish for free